Metaspace

Miguel A. Sanchez-Rey

Metaspace

By

Miguel A. Sanchez-Rey

Metamorphic space is defined as the following:

Cosmological homotopic states between variant [of stringy]'s.

These variant [of stringy]'s are said to be unique variations of one unique known variant call the Super Yang-Mills Gauge Analog [1]. All perturbations of such unique variant has led to the discovery of all but 32 variants that is to subsequently lead to the completion of the Grand Unification Scheme [1].

We can imagine each variant [of stringy] as encoding all the information of space-time from different variational parameters. Those parameters are set by the holographic counterterms.

It's clear now that:

Metaspace exist in D-energy meta-states [2].

D-energy is energy in D-variant. D-variant to be the extension of Dirichlet branes. Meta-states being different homotopic states.

It's postulated that there are more types of energy than the standard kinetic, potential, and etc., energies [including D-energy] already surmounted by current analysis. These are not forms of energy as nuclear, gravitational, and etc., by new types of energy that are subtler than the classical energy that is the bedrock of all known forms of energy [2]. The potential and kinetic energy being the limit of reducibility.

By extracting D-energy in metaspace we can manipulate the super-positional quantum state, such that, giving control of entropic and conservation of energy, entangle and disentangle molecules and atoms [2].

Finding other types of energy will be able to control the motion of matter such that there is no breakup of matter or even give way to different holographic transitory states of matter. So we can define as:

Metaspace exist in homotopic meta-states.

Homotopic meta-states being variational and perturbed.

The higher the energy-scale the more we gain access to metaspace. But metaspace will seem uncontrollable at first since new mathematical properties will be discerned and studied that it'll be an explosive phase.

Achieving control of metaspace means imposing SUPREME and Computational Control. Once SUPREME and Computational Control is imposed order will be established in metaspace that there's expected to be a beautiful tapestry of mathematical elegance and computational power that will reveal in higher-energy scales the completion of the Grand Unification Scheme [3].

References

[1] Sanchez-Rey, Miguel A. The Physicalist Program. Createspace: 2015.

[2] Sanchez-Rey, Miguel A. Meta-space Energy in the Grand Unification Scheme. Createspace: 2016.

[3] Sanchez-Rey, Miguel A. Energy-Scale of the Grand Unification Scheme. Vixra.org: 2016.

Explanatory Theory of Physics: The Grand Unification Scheme

Miguel A. Sanchez-Rey

Abstract

Why did unification work better with the Super-Yang Mills Gauge Analog than any other unification attempts?

July 19th, 2016

Explanatory Theory of Physics

The Super-Yang Mills Gauge Analog is said to be a variant [of stringy] [1]. This variant [of stringy] can be alternatively worded as a unification scheme. A variant [of stringy] is unlike what standard unification procedures expected. Grand Unification, at the time, sought to find a theory of everything Instead a grand unified theory has been developed. That is there are variants [of stringy] of perfect number that exist in metaspace and in which share similar variational parameters of holographic counterterms. These variants are unlike other unification schemes, in that, they are obligated to C2R or else they'll breakdown. These variants [of stringy] can be interrelated using the definition of grand unification scheme [2]. Other unification attempts did not yield discovery since they persisted on a fundamental theory of physics rather than an explanatory theory of physics. This explanatory theory of physics gave way to the discovery of the Super-Yang Mills Gauge Analog which led to the universal law of nature. But the universal law of nature is a crude method. It was realized that this variant [of stringy] yielded more variant [of stringy]'s so imposing the grand unification scheme, by renormalizing the Super-Yang Mills Gauge Analog, yielded a finite amount of variants of perfect number that can be related to each other through their charge monopoles.

References

[1] Sanchez-Rey, Miguel A. The Physicalist Program. Createspace: 2015.

[2] Sanchez-Rey, Miguel A. Physics in the Grand Unification Scheme. Vixra.org: 2016.

The Current Status of the Physicalist Program [PHPR]

By

Miguel A. Sanchez-Rey

The Physicalist Program [PHPR] was founded as a resolution to a foreseeable catastrophic scenario in the Scientific Age in the form of a task. One task is to pursued; then completed; before the next task is set. PHPR is to be dismantled when the last task is completed. By then the Scientific Age has come to an end...

The First Task of PHPR is a 100 year task that must be fully implemented when ITER [International Thermonuclear Experimental Reactor] goes online. Which then gives us a 40 year window of opportunity to complete 60 percent of the First Task of PHPR and 60 years to complete. The First Tasks confronts the question of how to resolve mineral depletion when ITER goes online and when ITER is mass manufactured 40 years after?

PHPR is now in the experimental stages. The 1st and 3rd task of the 7 Impossible Tasks has been partially completed but is expected to be fully completed when ITER goes online. The expectation of completion of those tasks yielded the conceptual and theoretical discovery of metaspace. Yet we still haven't gain access to metaspace. To gain access to metaspace requires an international effort of particle accelerators to go further up the energy-scale that will lead to the completion of the Grand Unification Scheme.

Pursuing metaspace, on a high-energy physics level, must await first conformation of supersymmetric particles at CERN, dark matter, and the achievement of ITER at a large-scale. But also it involves the abandonment of the hierarchy problem in favor of metaspace which means that evidence of supersymmetry will be found through metaspace.

Gaining access to metaspace will result in an explosive phase of mathematical discovery that will seem uncontrollable. In order to establish order in metaspace, at that point, we must impose Computational Control and SUPREME. But metaspace must be pursued carefully as a fatal misstep in metaspace will result in a transformative fall-out that gives little over 5 days before fatality.

The Second Task is yet to be decided but hints of that Task are beginning to emerge. The efforts of space-exploration, exo-biology, SETI [The Search for Extraterrestrial Intelligence], and the Kepler Satellite has increase and divulge more knowledge about life on other solar systems beyond ours. There may be, in the near future, evidence of life in our own solar system in the form of microbes. There may, as well, also be evidence of life even in the less inhospitable regions of the moons of Jupiter; complex life very much like planet Earth but very primitive and inexact.

Life may be observed, from a cosmic distance, in other planets or regions with new state-of-art satellite telescopes or we may be able to eavesdrop on other intelligent life-forms with new technology that takes advantage of sending digital information rather than detecting radio signals.

Completion of the First Task may pose a threat to an extraterrestrial intelligence. So the Second Task asks a possibly unique question. A question that involves the exhaustion of colonization and the development of space habitats within our own solar habitat. How do we proceed carefully beyond our solar system without becoming a threat to an extraterrestrial intelligence?

This conundrum relates to the apparent reality that colonization of other natural [or artificial] habitats is unviable since the biological nature of these planets do not correspond to the biological nature of Earth. Any attempt to colonize these planets may result in a full-scale confrontation, epidemic, or little time for habitation. In only the most extreme cases should the First Task of PHPR be applied for the development of habitats in an inhospitable moon or exo-planet that shows promise of a successful large-scale transformative reaction.

But yet the milestone of PHPR will be pursued to achieve space exploration which poses the question: how do we achieve transportation from one point in space-time to the other with little effort and little time? This is the development of star gates which is the most feasible transportation system since worm-holes are, theoretically, a naturally occurring phenomenon.

The Second Task of PHPR utilizes the milestone of PHPR. Since any other avenue is improbable as the development of warp drives, in itself, poses technological risks that make their invention unlikely and too costly.

It's in this sense that the current status of PHPR has been stated. Future efforts will be made that tests the limits of the types of energy that may exist through the discovery of D-energy in metaspace; the role of SUPREME in cosmology; and the theoretical and practical applications of computational control.

Entropy and Stability in the Grand Unification Scheme

Miguel A. Sanchez-Rey

Abstract

Does the grand unification work when entropy is cause by too much D-energy?

August 11th, 2016

If $[\ ] \xrightarrow{C2R} \Pi_\mu$ s.t. $p[n] \xrightarrow{C2R} [\ ]$ where $E = \blacktriangle$; then, $E = \blacktriangle + \blacktriangledown + ... \implies E = [\ ]$.

By imposing SUPREME we can control entropy as D-energy increases in metaspace [1]. As D-energy increases we head further into grand unification scheme where stability is kept optimal [2].

References

[1] Sanchez-Rey, Miguel A. D-variants and D-branes. Vixra.org: 2015.

[2] Sanchez-Rey, Miguel A. Energy Scale of the Grand Unification Scheme. Vixra.org: 2016.

www.ingramcontent.com/pod-product-compliance
Lightning Source LLC
Chambersburg PA
CBHW080535190526
45169CB00008B/3182